INFOGRAPHICS: AGRICULTURE

Enviro-Graphics Jr.

Renae Gilles

Published in the United States of America by:

CHERRY LAKE PRESS
2395 South Huron Parkway, Suite 200, Ann Arbor, Michigan 48104
www.cherrylakepress.com

Reading Adviser: Beth Walker Gambro, MS Ed., Reading Consultant, Yorkville, IL

Photo Credits: ©grafikacesky/Pixabay, cover; ©OpenClipart-Vectors/Pixabay, cover; ©Shutterstock, cover; ©Shutterstock, 1; ©Clker-Free-Vector-Images/Pixabay, 5; ©gustavofer/Pixabay, 5; ©OpenClipart-Vectors/Pixabay, 5; ©PinkPanthress/Pixabay, 5; ©Shutterstock, 5; ©GoodStudio/Shuttershock, 6; ©iStockphoto/Getty Images, 8; ©grafikacesky/Pixabay, 8; ©OpenClipart-Vectors/Pixabay, 8; ©iStockphoto/ Getty Images, 11; ©iStockphoto/Getty Images, 12; ©Shutterstock, 12; ©DigitalVision Vectors/Getty Images, 13; ©Clker-Free-Vector-Images/Pixabay, 14; ©Shutterstock, 14; ©Clker-Free-Vector-Images/Pixabay, 15; ©Janjf93/Pixabay, 15; ©mohamed_hassan/Pixabay, 15; ©Shutterstock, 15; ©Shutterstock, 16; ©Shutterstock, 18; ©Shutterstock, 19; ©Shutterstock, 20; ©Shutterstock, 21; ©DigitalVision/Getty Images, 22

Copyright ©2023 by Cherry Lake Publishing Group

All rights reserved. No part of this book may be reproduced or utilized in any form or by any means without written permission from the publisher.

Cherry Lake Press is an imprint of Cherry Lake Publishing Group.

Library of Congress Cataloging-in-Publication Data has been filed and is available at catalog.loc.gov

Cherry Lake Publishing Group would like to acknowledge the work of the Partnership for 21st Century Learning, a Network of Battelle for Kids. Please visit http://www.battelleforkids.org/networks/p21 for more information.

Printed in the United States of America
Corporate Graphics

ABOUT THE AUTHOR

Renae Gilles is an author, editor, and ecologist from the Pacific Northwest. She has a bachelor's degree in humanities from Evergreen State College and a master's in biology from Eastern Washington University. Renae and her husband live in Washington with their two daughters, Edith and Louisa.

CONTENTS

What Is Agriculture?	4
A World of Agriculture	6
Issues with Agriculture	10
Possible Solutions	18
Activity	22
Find Out More	23
Glossary	24
Index	24

WHAT IS AGRICULTURE?

Agriculture means farming. Taking care of the soil and growing crops are part of agriculture. So is raising animals. Crops and animals are needed to feed people around the world. Agriculture also provides many jobs. It makes money for people and countries. But some farming methods can be bad for the environment.

AGRICULTURAL PRODUCTS

Food from Plants

Fruits

Vegetables

Grains

Animal Products

Meat and fish

Dairy

Eggs

Materials

Cotton

Wool

Lumber

A WORLD OF AGRICULTURE

Organized agriculture began thousands of years ago. Over time, people have discovered new methods and technologies. Different countries focus on different products. These depend on the type of land and the weather of a place.

COUNTRIES WITH THE MOST AGRICULTURAL LAND

CHINA — 3,279,181 SQUARE MILES*

UNITED STATES — 2,521,913 SQUARE MILES

AUSTRALIA — 2,305,772 SQUARE MILES

BRAZIL — 1,761,873 SQUARE MILES

RUSSIA — 1,352,861 SQUARE MILES

*1 square mile = 2.6 square kilometers

2016, World Bank

TODAY'S U.S. FARMS

One farm feeds about **166** people.

About **25%** of all farmers are new to farming.

The top **3** farm products are cattle, corn, and soybeans.

About **36%** of farmers are women.

2017, U.S. Department of Agriculture

MOST VALUABLE AGRICULTURAL PRODUCTS BY STATE

 Fruit

 Vegetables

 Cow's milk

 Cattle

 Fish

 Flowers and greenhouse plants

 Other crops and hay

 Grains and beans

 Poultry and eggs

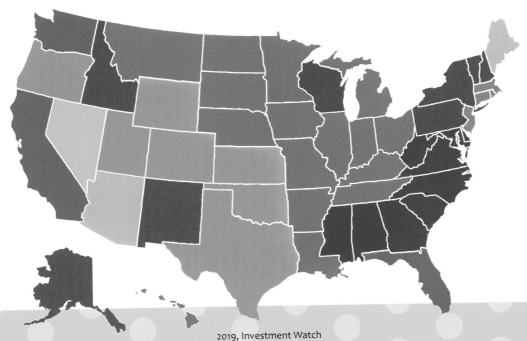

2019, Investment Watch

ISSUES WITH AGRICULTURE

Agriculture is needed for people to live. But it comes with many issues. Certain agricultural practices pollute Earth's water, land, and air.

Pesticides are sprayed onto crops to keep bugs away. Herbicides are used to kill weeds. These chemicals protect crops. But they can also harm the environment. They are not only toxic to the plants and animals they are trying to keep away. Many are also toxic to other animals, plants, and humans.

FAST FACTS

About **33%** of human food is lost or wasted every year.

Each person in Europe and North America wastes **209** to **254** pounds (95 to 115 kilograms) of food a year.

Every year, **$1 TRILLION** worth of food is wasted.

HOW FOOD IS WASTED

Production
- Unharvested crops
- Death from disease

Grocery stores
- Overstocking
- Selection for ideal shape and size

Processing
- Bad storage conditions
- Overproduction

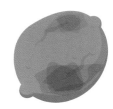

Home
- Oversized portions
- Uneaten leftovers

2015, FAO

GLOBAL FOOD WASTE

Fruits and Vegetables
(equal to 3.7 trillion apples)

Cereals and Grains
(equal to 763 billion boxes of pasta)

Dairy
(equal to 574 billion eggs)

Fish and Seafood
(equal to 3 billion Atlantic salmon)

Meat
(equal to 75 million cows)

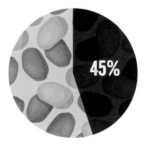

Root Vegetables
(equal to 1 billion bags of potatoes)

2012, Food and Agriculture Organization of the United Nations

LAND ISSUES

Habitat Loss

As of 2019, **1/2** of habitable land goes to agriculture.

More than **5,000 species** are threatened by agriculture.

Most habitat loss comes from livestock production. As of 2020, livestock use **77%** of the world's farming land.

Deforestation

In 2017, it was estimated that all rainforests would be cleared in **100 YEARS**.

As of 2013, about **80%** of **deforestation** in the Amazon was for cattle farming.

Humans have cut down **46%** of the world's trees.

WATER AND AIR ISSUES

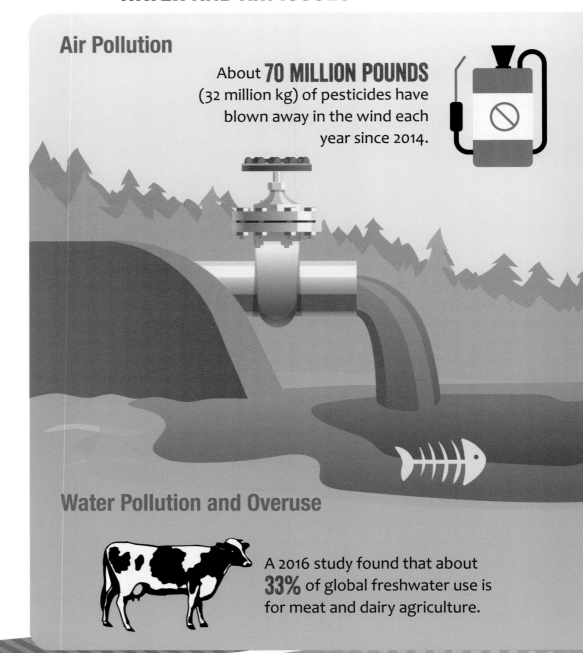

Air Pollution

About **70 MILLION POUNDS** (32 million kg) of pesticides have blown away in the wind each year since 2014.

Water Pollution and Overuse

A 2016 study found that about **33%** of global freshwater use is for meat and dairy agriculture.

Agriculture is responsible for **16%** of one dangerous type of air pollution, said a 2009 study.

Since the early 2000s, U.S. food travels an average of **1,500 MILES** (2,414 kilometers) by truck, boat, and plane, which all put exhaust into the air.

As of 2019, meat and dairy cause **57%** of global water pollution.

Fertilizers and pesticides have made more than **400** ocean "dead zones," areas where normal sea life cannot exist.

SYNTHETIC VS. NATURAL PESTICIDES

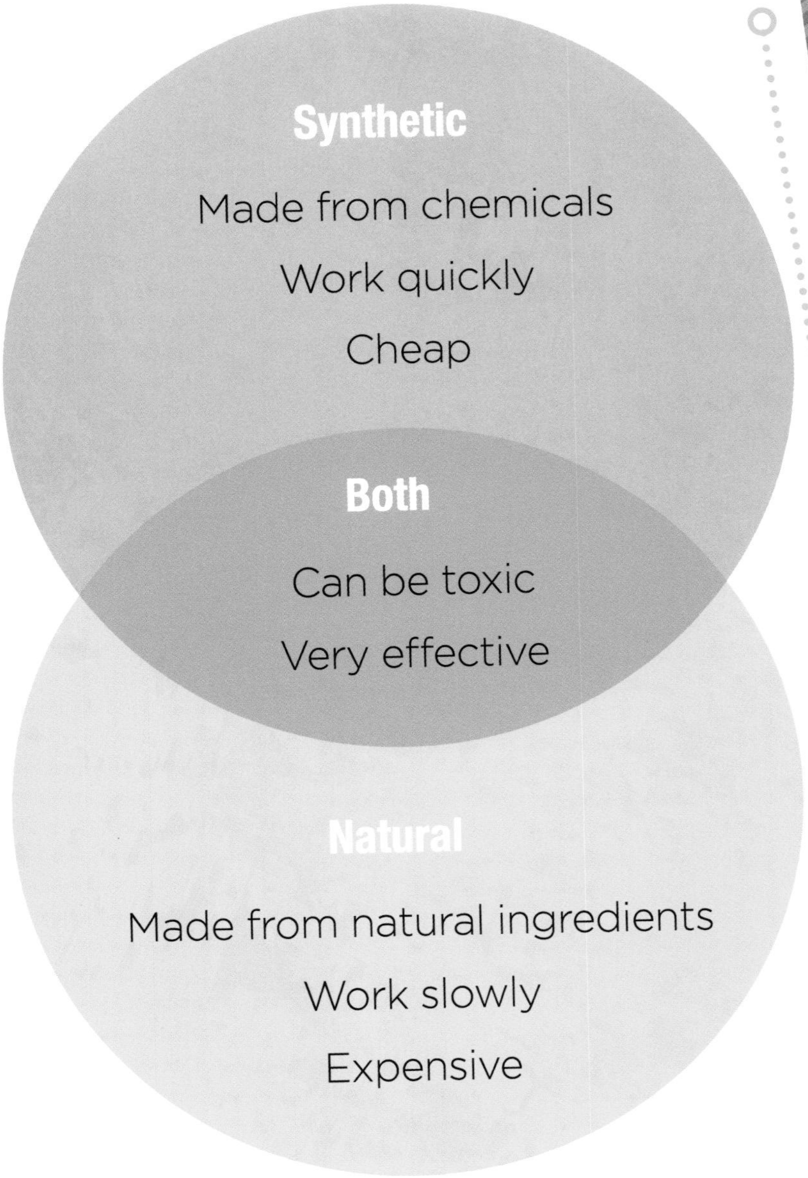

Synthetic
- Made from chemicals
- Work quickly
- Cheap

Both
- Can be toxic
- Very effective

Natural
- Made from natural ingredients
- Work slowly
- Expensive

POSSIBLE SOLUTIONS

As people learn more about agricultural issues, they come up with solutions. There are many things farmers and other people can do to help.

Composting is one way to reduce food waste. Food scraps and certain leftovers can be put to good use. People can compost at home. Many **municipal** areas have composting programs too.

SOLUTIONS

What You Can Do

- **Reduce food waste.** Several U.S. agencies have a goal to reduce food waste to 109 pounds (49 kg) per person a year by 2030.
- **Eat less processed food.** Food processing creates pollution.
- **Buy local.** This reduces food waste and pollution.

What Farmers Can Do

- **Use fertilizers and pesticides carefully.** Proper timing can reduce the amount of chemicals needed.
- **Plant trees and shrubs around fields.** This can lead to 20% more yield in certain crops.
- **Feed livestock quality food.** Grazing cows produce 50% less methane.

FAST FACTS

Since 2017 in the United States, **94%** of food scraps end up in landfills.

As of 2017, food and yard waste make up about **28%** of a household's garbage.

Composting can lead to **LESS GARBAGE**.

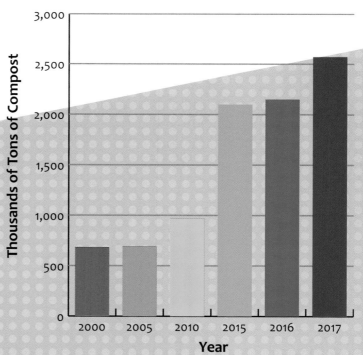

2019, U.S. Environmental Protection Agency

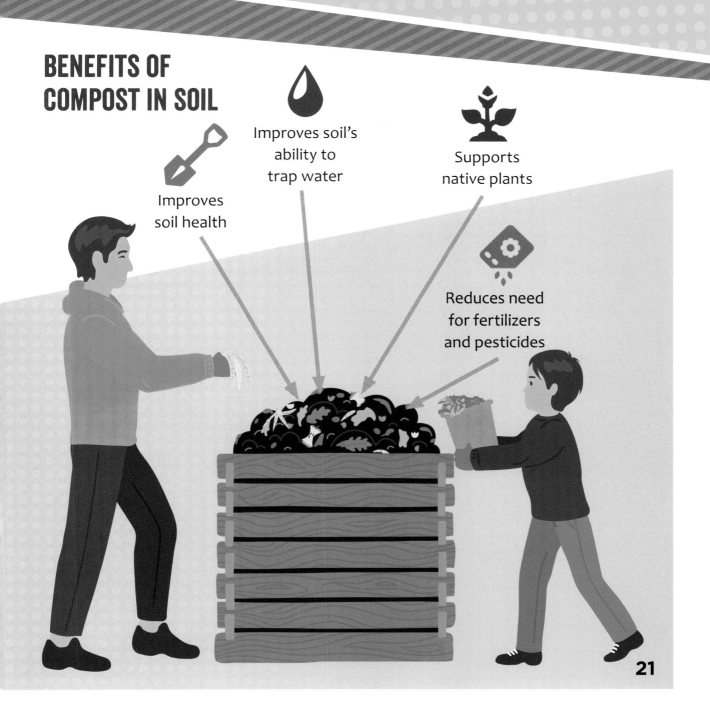

ACTIVITY

Grow a Veggie Garden

You can get involved in agriculture by growing your own vegetable garden.

1. Research what veggies grow well in your area. Lettuce, carrots, and radishes are usually good for beginners. You can pick out seeds from a local store or check to see if local gardening groups have seeds for free.

2. Write out a plan for your garden. When should you plant your seeds? Work with a parent or guardian to set up an area outside or a planter inside your home.

3. Plant your garden and keep track of its progress. If you have a problem with bugs, research natural ways to solve it.

4. Enjoy the fruits of your labor!

FIND OUT MORE

Books

Johnson, Robin. *Food Scientists in Action.* New York, NY: Crabtree Publishing, 2019.

Labrecque, Ellen. *Permaculture.* Ann Arbor, MI: Cherry Lake Publishing, 2018.

Smundak, Katharina. *The Truth Behind GMOs.* New York, NY: Rosen Central, 2018.

Websites

Action for Healthy Kids
https://www.actionforhealthykids.org/activity/food-waste

Britannica Kids: Agriculture
https://kids.britannica.com/kids/article/agriculture/352715

NeoK12 Agriculture
https://www.neok12.com/Agriculture.htm

GLOSSARY

composting (KOM-pohst-ing) using decaying material, such as leaves or vegetable scraps, to improve garden soil

deforestation (di-for-eh-STAY-shuhn) the removal of all the trees in an area

municipal (myoo-NISS-uh-pul) having to do with the government of a city or town

species (SPEE-seez) a group of animals or plants that share similar traits

yield (YEELD) how much a particular farm, field, or plant produces

INDEX

air, 10, 14, 15

chemicals, 10, 17, 19
composting, 18, 20
crops, 4, 9, 10, 11, 19

emissions, 19

farming, 4, 8, 13
fertilizers, 15, 19, 21
food, 5, 11, 12, 15, 18, 19, 20

herbicides, 10

land, 6, 7, 10, 13

pesticides, 10, 14, 17, 19, 21
pollution, 14, 15, 19

soil, 4, 21

waste, 12, 18, 19, 20
water, 10, 14, 15, 21